Concepts of Sexual Health Sex & You!

Leader's Guide
(Modified for Jr. High)

Concepts of Truth, Inc.
PO Box 1438
Wynne, Arkansas
info@conceptsoftruth.org
870.238.4329

ISBN: 978-0-9849652-8-1

©2015 Revised

©2018 Revised

©2019 Modified for Jr. High

Concepts of Sexual Health Sex & You!

(Modified for Jr. High)
TABLE OF CONTENTS

The purpose of the curriculum is to teach that the human person exists as a multidimensional being and that one's sexuality is integrated in all the dimensions. As human beings, we have the capacity to reason, make choices, seek what is true, and to ask questions of ultimate significance.

The lessons in the curriculum stress that one's sexual health depends on the choice to save sex for marriage or one monogamous bond and by making healthy choices in relationships, love, and responsibility. These choices affect the whole person's present and future well-being and the heritage passed on to future generations.

According to the Arkansas State Health profile (at time of print) (insert applicable state/country's STD stats) http://www.cdc.gov/nchhstp/stateprofiles/pdf/arkansas_profile.pdf, Arkansas ranks eighth for Chlamydia and sixth for Gonorrheal infections in the United States.[1] Since sex involves one's sexuality, is more than an isolated act, is more than a bodily function, is not just a commodity to buy or sell, and since contraceptives do not always provide protection against STDs, sexual health education must appeal to the dispositions of the mind, heart, will, and conscience of the learner. Young people can be reached through their hearts and challenged to become persons of character, capable of contributing not only to their own well-being but also to their communities and society.

The Whole Person Learning Theory by Onalee McGraw, Ph.D., founder of the Educational Guidance Institute, Inc. and author of *Teaching the Whole Person about Love, Sex, and Marriage Educating for Character in the Common World of our Homes, Schools, and Communities*, explains the philosophical and psychological concepts of the whole person approach to learning: the philosophy that one's sexuality is an integral part of the whole person. *"As whole persons, men and women possess a sexuality that is integrated in body, mind, heart, will, and conscience. The sexual domain is permeated by the intellectual, moral, emotional, physical, and social domains. The psychological guiding principle is that cognitive powers of memory, thought, judgment, imagination, and learning related to sexuality permeate the body, mind, heart, will, and conscience; thinking, feeling, and acting in the sexual domain reflect the whole person."*[2] The premise of the whole person learning theory is foundational for *Concepts of Sexual Health Sex & You!*

The resources included in *Concepts of Sexual Health Sex & You!* (modified for Jr. High) are based on medical facts, statistics from the Center for Disease Control & Prevention/cdc.gov, and are developmentally appropriate for 6-8 grades. The curriculum also meets Arkansas' 6th, 7th and 8th Grade Health & Wellness content standards. See student expectations listed for the various lessons.[3] Permission is given by the authors to insert applicable substitutes for video clip(s), handouts, or slides based on current stats/population/culture maintaining the meaning and values of the original content.

In five 45 minute lessons using video, PowerPoint, lecture and discussion, *Concepts of Sexual Health Sex & You!* presents Jr. High students with the concepts of the whole person approach and a definition for sexual health that integrates the five dimensions of the whole person; human dignity and development; promotes virtue and character in love and marriage; medically based facts about STDs, contraception, and prevention of disease; and practical help to guide young people in the present and future when seeking answers to life's most difficult questions. Acquiring this knowledge and life skills will give students the opportunity to pass on sexual health and a heritage for life to future generations.

For the Teacher Lesson I
Concepts Sexual Health Sex & You!

The lesson begins with building rapport with students. Josh McDowell says, *"Rules without relationships = rebellion."* [4] Building relationships with students will produce more positive outcomes with the curriculum. Students do not care how much you know until they know how much you care. Show your passion for the subject matter! Mutual respect opens the door to hearts and lives. The lesson gives the student a glance at their preferred future and the kind of heritage they may want to pass down to future generations. Applying the five interactive dimensions of the whole person - physical, social, emotional, intellectual, and moral, students will gain knowledge and understanding of their sexual health lived out in true freedom with purity, integrity, and unconditional love. The lesson also helps students to consider long term academic, vocational, and social goals, which are a critical factor in helping adolescents avoid negative pressures toward sexual activity outside of marriage or one monogamous bond. *Concepts of Sexual Health Sex & You!* lessons help students consider their life's purpose, goals and to consider their complex whole being by making healthy choices in relationships, love, and responsibility.

Lesson I Objectives:

1) Teachers and students will develop a rapport by getting acquainted allowing each one to tell their name and a few interests of sports, hobbies, etc. with at least 90% participation from class.
2) Participants will be able to define sexual health and describe the purpose of sex in relation to the five interactive dimensions of the whole person, i.e., physical, emotional, intellectual, social and moral with 75% participation.
3) By comparing risky sexual behaviors outside of one monogamous bond, students will be able to list benefits of integrity, purity, and unconditional love within the five interactive dimensions of the whole person. During lecture, Ppt. presentation and discussion, students will give an acceptable range of answers that are allowable as correct.

Lesson Plan:

Permission is given by the authors to insert applicable substitutes for video clip(s), handouts, or slides based on current stats/population/culture maintaining the meaning and values of the original content.

Activities: Sonic Coupon Activity, Whole Person Glove Activity, Multi-Color Gummy Worm Stretch, Duct Tape/Bonding

Lecture & Discussion

Video: *A Million Ways to Say No : Developing the Whole Person*

PowerPoints (Ppt.): (1) A Million Ways to Say No (2) Choices, Choices (3) The World Health Definition (4) COTI's Sexual Health Definition (5) The Five Aspects of the Whole Person (6) Sex & You (7) Question (8) Whole Purpose of Sex (9) The Gift of Self (10) Welcome Back (11) The Five Aspects (12) Developing the Whole Person Video (13) The Five Aspects (14) COTI's Sexual Health Definition (15) Science of Sex (16) Duct Tape Activity (17) Planning the Future (18) Planning Your Children's Future (19) Uniquely Different

Get Acquainted

Hello, my name is. . . from (local organization.) I'm excited to be here and I look forward to getting to know you better. There is only one rule. . . I will respect you and I expect you to respect me. Let's go around quickly and tell us your name and maybe something you enjoy. I will start. My name is. . . (leader tells name and what they enjoy.)

Introduction

What is this class all about? Sex! What? Ugh. . . (Shake your head, grit your teeth, hold your head on each side, etc.) Yes, and I am called the "sex lady"! (Idea for two presenters one to ask and one to answer.)

This class will help you consider the five aspects of a whole person while making healthy choices. What is that? Are you a whole person? Absolutely – and you have 5 various aspects of your person. The name of this class is *Concepts of Sexual Health Sex & You!* This is not a "**Just say no to sex**" abstinence program.

Let's watch a **video**.

Ppt. 1, Lesson 1

Play (attention getter) Video: *A Million Ways to Say No,*

(Laser Time, 2016 https://youtu.be/N3sb_Iw7Zyo) (Works well if you ask them to see if they can name some of the characters.)

It **IS** good to say **NO** to bad things like drugs, But **SEX is GOOD** with one monogamous (means 1 and only) partner and hopefully in the bonds of marriage. So, this class is not **just say no to sex**, it is about your sexuality being interconnected to your whole person. It is about finding sexual health through purity and choosing to save sex for that one monogamous bond. The class is **not merely about saying "no" to sexual activity but about saying "yes" to love and responsibility**. Sexual Health is about **YOU** - your whole self.

Sonic Coupon Activity: (any restaurant coupon or candy will work as reward)

How many of you would choose making lots of money over being in poverty? How many would choose a nice home over being homeless? How many of you would want good things for your children? I think basically, we all want good things. Right? How many of you would like a good thing like a free sonic coupon? Remember when choosing sides for a team and wanting to be chosen for a certain team game or sport? Yes, we all want to be chosen and I need to make a choice here. (As hands are raised, leader will take some time in choosing a recipient. Walk around, scratch your head, maybe have a couple to stand up and compare shirts, shoes etc. Tell them you like or dislike something about them and finally choose one or more to receive a coupon and tell the others you are sorry they were not included in your choice.)

We make choices every day and **choices have consequences!** Some consequences are good, and some are bad. As a leader, I just made a choice as to who would receive these free coupons and the consequence happened to be a good one. . . someone won!

We all want to be chosen and as human persons we all have the dignity and deserve the right to be chosen! We all have a need to be liked, included, and desired. We value good things. Down deep in our inner core is the natural human desire to give and receive unconditional love. We long for love, relationships and of course. . .free things. . .like these sonic coupons! (Give away one or more coupons.)

The things we desire and can choose in life are both tangible (can see) and intangible (cannot see). They include things like someone to love us unconditionally, a fairy tale romance, good jobs, nice homes, adventure, happiness, peace, and contentment.

Unfortunately, our choices do not always produce the desired outcome or positive consequences. Sometimes in life we think we are headed in the right direction, only to find the choice we took led us to pain and heartbreak.

Do you know anyone who is divorced? Some people do not get married at all - Some like friends with benefits or hookups. They just cruise from one person to another – one bed to another. These are people who seem to make poor choices and end up getting hurt.

The next PowerPoint shows a person trying to choose. . .if only we could look ahead and see inside the treasure chest!

Ppt. 2, Lesson 1

(Some consequences are evident and others are harder to see.)

Is the romance of Cinderella and her Prince just a fairy tale, or is it possible to find the one true love of your life. . .and keep them? Fortunately, there is! There is a way to find true love, true freedom, sexual purity, and a way to make healthy choices in relationships, love, and responsibility.

You can have sexual health as a whole person and know a relationship based on unconditional, committed married love. Author Onalee McGraw, Ph.D [5] says that this can be achieved by *"all persons through their natural human capacities of reason, desire for the good, voluntary will and moral sense."* [6]

She says this can *"be true for all persons regardless of their own particular family background, the community where they live, or the cultural era in which they come of age."* [7]

What is Sexual Health?

The World Health Organization says, *" sexual health is not merely the absence of disease, but it requires a positive and respectful approach to sexuality and sexual relationships, as well as safe experiences, free of coercion and violence."* [8]

Ppt. 3, Lesson 1

The *World Health Organization* says…

> **"Sexual health is not merely the absence of disease, but it requires a positive and respectful approach to sexuality and sexual relationships, as well as safe experiences, free of coercion and violence."**

According to research on sexually transmitted diseases and the fact that one's sexuality is interconnected and integrated (**hold up your hand and wiggle fingers**) throughout the whole person, we cannot have a safe sexual experience except in the bonds of one monogamous partner expressing unconditional love.

Our sexual health depends upon living out our sexuality in true freedom and wholeness, which involves the five dimensions or aspects of the whole person. Let's learn a definition for sexual health based on the whole person concept.

Ppt. 4, Lesson 1

(Concepts of Truth's Definition)
Sexual Health is.........

living out one's sexuality in true freedom with integrity, purity and unconditional love.

It is making healthy choices in relationships, love and responsibility that will affect the whole person present and future and the heritage passed on to future generations.

Y❤U

Sexual health is living out one's sexuality in true freedom with integrity, purity and unconditional love. It is making healthy choices in relationships, love, and responsibility that will affect the whole person present and future and the heritage passed on to future generations.

Anyone who can learn Concepts of Truth's Sexual Health definition word for word by the end of our classes will be eligible to be in a drawing for a gift! Now, let's read the definition again. (Read definition.)

How would you describe unconditional love?

To have true freedom, integrity and purity in sexual health means to consistently live our lives with the highest moral sexual standards and unconditional love consistently guarding your mind, will and emotions from sexual impurity. To live sexually healthy **is to be the same in the dark as you are in the light**. To have sexual health means we set limits on how much of our bodies we share with others. Sexual health is possible for anyone, regardless of past choices or painful experiences.

Think about the following question:

How could your sexual health affect the goals you have for your life?

Sex & You!

Sex is not an isolated act. It is not about **doing** but it is about **being** male or female. Sex is you - the whole person! Our sexuality is interconnected and integrated throughout the whole person and affects us intellectually, morally, emotionally, physically and socially. Sex is more than a bodily function and not just a commodity to buy or sell.

Contraceptives like birth control pills and condoms do not always provide protection against STDs so sexual health education must appeal to dispositions of the mind, heart, will, and conscience of the learner. Author Onalee McGraw, PhD describes **The Five Aspects of the Whole Person.**

(Co-leader interrupts with the following whole person glove activity. The purpose of this exercise is to reinforce the five aspects of the whole person while engaging and capitalizing on the imagination and assumptions of the learner.)

Whole Person Glove Activity

Preparation:
(1) Prepare a latex glove in advance by writing the five aspects of the whole person on each finger of the glove with a permanent marker.

(2) Fold the glove lengthwise onto itself until it slightly resembles a condom, then roll the glove up from the end to the opening until it is compact and can easily slip into a pocket for retrieval.

(Co-leader interrupts)
"You know… I brought something with me today that I tend to use all the time. Maybe you've seen one of these around the house… or maybe your parents use these." (As you are talking begin to pull the glove out of your pocket, but keeping it contained as you speak. Done well, this will cause assumptions to occur with whispering and giggling among the students.)

"What I really like to do with these, is to blow them up like balloons." (Take the glove, blow it up like a balloon to expose the 5 aspects of the whole person written on the fingers. Students will realize they made the wrong assumption, while allowing a humorous and engaging transition to the discussion of the 5 aspects of the whole person.) "The 5 aspects of the whole person - Intellectual, Moral, Physical, Emotional, and Social!"

(Teacher continues)
As I was saying, Dr. McGraw describes **The Five Aspects of the Whole Person** and says, *"Life is experienced on many levels. In our attitudes, relationships and behavior we all experience life as whole persons. In human life, our reason, moral sense and emotions, our bodies and our social relationships are all in play simultaneously. Furthermore, each of these domains is affected by the others. The human person must always be considered as a whole person - never in parts."* [9]

Ppt. 5, Lesson 1

The Whole Person

Intellectual
Human persons possess an intellect enabling them to think and grasp abstract ideas, and a will enabling them to make rational choices.

↓

Moral
Our intellects enable us to recognize the distinction between good and evil, right, and wrong. Our will is then confronted with the task of living accordingly.

↓   ↓

Emotions and Feelings
Emotions and feelings permeate our whole person. Mature persons have the ability to control their emotions, subjecting them to intellectual and moral standards.

Physical
The body is the living organism that houses each of us. Ideally, our physical actions will be controlled by properly ordered emotional responses, but since the body sometimes matures more quickly than the intellect and the emotions, this is not always easy.

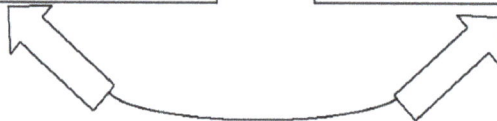

Social

Each person is part of a community. As we grow, we learn to interact with others, develop friendships and work to benefit others as well as ourselves.[10]

(Example: Using the emotion of anger in reaction to the aspects works as an excellent associative sequel into the challenge expressed in the "physical" box. It seems to resonate well with Middle Schoolers. Tapping into their existing experiences is essential.)

Let's remember the whole person on our five fingers. The five aspects are intellectual, moral, emotional, physical, and social. Give your neighbor a "high five."

Ppt. 6, Lesson 1

Sex and YOU!

The human person is uniquely intricate and whole. Our body, mind, heart, will, emotions, and our social and moral sense function interactively to make us who we are. Our sexuality is an integral part of our entire being and our unique and personal inner core. In this inner core is the natural human desire to give and receive unconditional love.

The human person is uniquely intricate and whole. Our body, mind, heart, will, emotions, and our social and moral senses function together and are integrated to make us who we are.

Our sexuality is an integral part of our entire being and our unique and personal inner core. In this inner core is the natural human desire to give and receive unconditional love. So, now we have the title of this program...*Concepts of Sexual Health Sex & You!*

Activity: Object Lesson – Multi-color Gummy Worm Stretch – (Ask for 2 volunteers and give each one a gummy worm. Use multi-colored ones and discuss how it looks as if the gummy worm is made up of different segments or parts. They may or may not have 5 – but some do have five variations of colors.) (We have found it works well to allow all to participate.)

As we have discussed, the human person is intricate and whole. Our intellect, moral sense, emotions, physical, and social aspects make us who we are. We all are growing and developing. Sometimes Jr Highers may even feel those changes and feel confused with physical changes, emotions, social relationships, and so on. And just as we can stretch this candy – our whole self can feel stretched at times. Let's see which one of you can stretch your gummy worm the farthest. I have a sonic coupon for the winner!

If Sex is Not an Act, What is the Purpose?

Ppt. 7, Lesson 1

YOU

If sex is not merely an act, what is the purpose

?

Ppt. 8, Lesson 1

y X U   WHOLE
              PURPOSE OF
                  SEX

                Procreation
          Bonding/Unitive Intimacy
                Completion
             Marital & Family Life
Source: J. Budziszewski : What We Loose When We Forget What Sex Is For"

The WHOLE Purpose of Sex
Sex is for procreation, bonding/unitive intimacy, completion, marital & family life... .and yes, it is pleasurable!
Sex between a male and female (complementary opposites) brings a man and woman together in unitive intimacy. Author J. Budziszewski says, *"...the longing for unitive intimacy is at the center of our design."* [11]

In the research study report, *"Hardwired to Connect: The New Scientific Case for Authoritative Communities."* 2003 issued by the Commission on Children at Risk, the authors state there are two kinds of connectedness: close connections to other people, and deep connections to moral and spiritual meaning. Plank #10 states, *"The human brain appears to be organized to ask ultimate questions and seek ultimate answers."* [12]

We all long for the purpose of sex - transcendent, unconditional love, connecting to the opposite sex and producing children as a symbol for this union of completeness.

Pleasure is a byproduct of sex, but not the ultimate goal. Pleasure comes naturally as a byproduct of pursuing something else. For instance, the purpose of eating is for nutrition, but eating is pleasurable. If we ate simply for the pleasure it would be unhealthy. If we use sex simply for pleasure, someone gets used. It becomes selfish and disordered.

Sex is for procreation (producing children), **unity, completeness, and family life.** And outside of celibacy (abstaining from sexual intercourse, especially by reason of religious vows, unmarried), there is something missing in the man, which must be provided by the woman and vice versa. **Men and women complement each other.** The book *Men are Like Waffles and Women are Like Spaghetti* [13] is a good resource to explain the complementary relationship.

There is a **"natural law of sex"**.[14] Physically, in order to procreate a man and woman need each other. But also, we long for the unitive intimacy in a relationship based on unconditional love.

Marriage and family life are part of natural law. There is never a time in human history when they did not exist. According to the Universal Declaration of Human Rights, the family is the natural and fundamental group unit of society and is entitled to protection by society and the State.[15] We are not designed for "hooking up," but we are designed for our bodies and our hearts to work together.

The ideology of hooking up says that sex is merely release or recreation. Author J. Budziszewski goes on to say, *"Mutual and total self-giving strong feelings of attachment, intense pleasure, and the procreation of new life are linked by human nature in a single complex of purpose. If it is true that they are linked by human nature, then if we try to split them apart, we split ourselves."* [16]

Budziszewski says:
"The gift of self makes each self to the other-self what no other self can be".[17]
(Have students try saying it faster just for fun and to help them remember the phrase.)

Ppt. 9, Lesson 1

"The gift of _self_...   ¥ ⚡U
    makes each _self_...
        to the other _self_...

    what no other self can be."

To "forsake all others" is not just a sentimental feature of traditional Western marriage vows; it arises from the very nature of the gift. You cannot partly give yourself, because yourself is indivisible as a whole person; **the only way to give yourself is to give yourself entirely.**

Because the gift is total, it has to exclude all others, and if it doesn't do that, then it hasn't taken place.[18] Sex involves the whole person. Our sexuality is a gift to be given with unconditional love. It is a whole act with a **WHOLE PURPOSE.**

(Note: 45-minute class time ends here.)

Ppt. 10, Lesson 1   **Ppt. 11, Lesson 1**

Welcome BACK! The last time we met, we discussed you as a whole person with five aspects. I have a free coupon for someone who can name those five aspects. (Intellectual, Moral, Physical, Emotional, and Social). Let's watch this short video about young people developing as a whole person. As you watch, see if you can discover examples of the five aspects.

Ppt. 12 (Play Video: Developing the Whole Person) [19]
https://youtu.be/VDFonhPChmY

Which one of the five aspects was not mentioned in the video?

The video did not mention the emotional aspect. The emotional aspect of the whole person includes our feelings: Sadness, joy, anger, disgust, love, forgiveness. Last week we mentioned there is something we all desire. What is something we all desire? To give and receive unconditional love.

Ppt. 13, Lesson 1

Concepts of Truth's definition of sexual health says:

Sexual health is living out one's sexuality in true freedom with integrity, purity and unconditional love. It is making healthy choices in relationships, love, and responsibility that will affect the whole person present and future, and the heritage passed on to future generations.

Ppt. 14, Lesson 1

We also learned last week that sex is not just about "doing" an act, but it is about our "being" male or female. When we have a sexual relationship with someone, the bonding is not just physical. It affects the whole person. What we are really looking for when having sex outside of marriage is real intimacy - interpersonal communication at the deepest level.

We look to sex to provide the closeness and love that we're longing for. The bonding happens whether one is married or not. "Sex makes us feel close even when we hardly know each other. Yet couples who initiate sex early on in a relationship have difficulty moving to that deepest level and experiencing true intimacy with each other." [20]

The Science of Sex
Ppt. 15,   Lesson 1

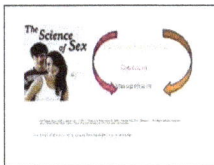

A "science of sex" occurs when bonding hormones oxytocin and vasopressin are released during the act of intercourse.[21] This bonding is not outwardly seen but inwardly expressed.

During sexual activity, these powerful hormones are released in the brains of men and women that produce lasting bonds with their partner. Oxytocin is a bonding hormone released during childbirth and nursing that causes the mother to bond with her infant.

It is also released during sexual activity and acts as emotional super glue between partners. Both men and women have oxytocin and release it during sexual activity; but women are more affected by oxytocin and men by vasopressin, another bonding hormone released during sex.[22]

Vasopressin helps a man bond to his partner and instills a protective instinct toward his partner and children.

Ppt 16, Lesson 1 Duct

Tape Activity

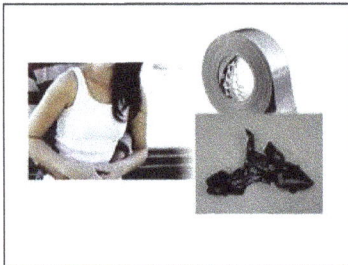

(If you have a male and female willing to participate, wrap duct tape around one of their arms and then the arm of the other so they are stuck together. You can also wrap duct tape around your own arm to demonstrate how bonding takes place.)

When a couple has sex, the bonding hormones are released, so the tape is tight. Sex is like adhesive. Promiscuity is like taking the tape off again and again. We want to be bonded forever with one monogamous partner to live happily ever after.

(*Now take the tape off and try to stick the same tape a second time.*)

This represents the couple deciding to have different sexual partners. There are pieces of skin cells, hair, etc. on the tape, and it won't stick as good the second time without problems. **The choice to have more than a permanent, monogamous relationship creates problems intellectually, morally, emotionally, physically, and socially.**

Research suggests the ability to bond and produce oxytocin is damaged by the stress hormones released during a breakup. Just like debris on duct tape, previous sexual experiences reduce the ability to bond correctly. Oxytocin levels can return to normal if sexual activity is stopped, and time is given to address physical and emotional healing. **Refrain from getting into a new relationship for a year or two and commit to saving sex for marriage.**

Imagine the duct tape was never removed. The duct tape would begin to feel like a part of the arm, and the adhesion would be strong. **When a couple waits until marriage to have sex and remains faithful to each other during marriage, oxytocin, and vasopressin increase the biological bond between the husband and wife.**[23]

Remember... the definition of sexual health includes true freedom that will affect the future! Having sexual health is for a lifetime! It continues even after marriage. So, what is the purpose of sex? Procreation, Bonding, Completion, Marital & Family Life.

Planning the Future

Have you chosen to have sex before marriage? This choice allows risks in all areas of your life because you are a whole person.

Couples who cohabitate (live together before marriage) also increase their risks of damage to their sexual health. Let's read the risks listed in the Intellectual, Moral, Emotional, Physical, and Social dimensions: [24]

Ppt. 17, Lesson 1

(Note: Intellectual, Moral, Emotional, Physical, and Social will come up individually on the PowerPoint with examples. This PowerPoint and the next one are important parts of the first lesson. Take time to discuss each part.)

Planning Your Children's Future

If you have chosen to have premarital sex, that choice will not only affect your future but will also have an impact on your children's future. What kind of future do you want for them? Read what your children may experience in the five dimensions of the whole person. [24a]

Ppt. 18, Lesson 1

(Note: Emotional, Intellectual, Moral, Physical, and Social will come up individually on the PowerPoint with examples. This PowerPoint is an important part of this first lesson. Take time to discuss each part.)

Question:

Think about your recent choices. How have your choices affected your whole person?

Ppt. 19, Lesson 1

As a
WHOLE
Person

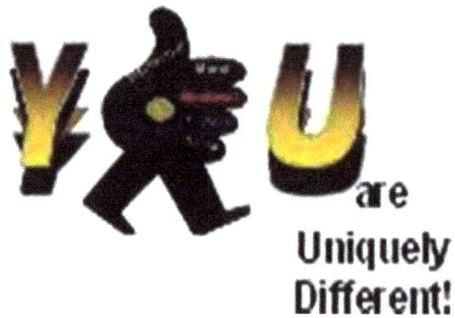

YOU
are
Uniquely
Different!

Uniquely Different! There is no one in the world, just like you! Your DNA, fingerprints, your hand clap, the pitch of your voice are all unique. Remember, men and women are uniquely different, not only as male, and female but as our individual self. Male and female are unique individuals with different roles that complement the other.
You are a whole person, and you are unique!

Note: We have found using personal testimonies of our unique personal value here has been very impactful on the students and allows them to understand that they cannot be replaced, and they are very valuable.

For the Teacher Lesson 2
Development & Dignity

In this lesson, students will explore fertility, prenatal development, and the dignity of all human persons. All human life has value, and our fertility gives us the power to create human life. The more adolescents understand the development and unique dignity of a human person, the intrinsic value of life, and the gift of fertility, the more they will be empowered in relationships to make healthy reproductive choices.

They will examine the premise that our bodies and our sexuality are an integral part of our unique and personal inner core. In this inner core is the natural human desire to give and receive unconditional love. Using video and models, they will view prenatal development and analyze the human life cycle from fertilization to infancy and their ability to create human life.

Objectives:
1) Students will compare and contrast male and female reproductive organs exploring the power of fertility and be able to describe that responsibility in sexual decisions, with 75% of students participating in the discussion.
2) Students will view the video, *9 Months in the Womb,* and a set of pre-natal models representing human development from 7- 30 weeks gestation. Students will be able to identify the stages of human development and describe the characteristics of each stage using a class discussion with at least 75% accuracy.
3) Students will be able to define intrinsic and extrinsic value and apply the definition to love and responsibility in relationships sexuality with 75% participation and accuracy.

Lesson Plan
 Permission is given by the authors to insert appropriate substitutes for video clip(s), handouts, or slides based on current stats/population/culture, maintaining the meaning and values of the original content.

Activity: **$20 Bill** (Giving Value & Dignity to A Human Person)

Video: *9 Months in the Womb* - A Remarkable Look at Fetal Development Through Ultrasound by PregnancyChat.com[25]

Touch of Life Fetal Models

Lecture & Discussion

PowerPoints (Ppt): (1) Male & Female Fertility (2) Horton (3) 9 Months in the Womb video (4) Question (5) The Dignity of a Human Person (6) Outward & Inward Dignity (7) Unconditional Love (8) Whole Person

Introduction:

Fertility, the Beginning

Today, we will explore the power of fertility, prenatal development, and the dignity of all human persons. **How does sex result in procreation?
What does it mean to be a whole person? Does human life have dignity (a presence that commands respect) and value?** Let's read the following ppt.:

Ppt. 1, Lesson 2

Our fertility gives us the power to create human life. Fertility is the state of or being fertile and fertile means; one is capable of reproducing. The male and female are complementary to each other. Females have open reproductive systems, and males have closed. The female provides the egg (ovum) with 23 chromosomes, the male fertilizes with his sperm of 23 chromosomes, and the female incubates the resulting zygote with 46 chromosomes (Human DNA genetic code).

Procreation also requires an enduring partnership between two beings. Why would both parents be needed? To raise the child, both are needed because the male is innately the protector and the female the nurturer. A parent of each sex is necessary to make the child, to raise the child, and to teach the child.

The child needs a model of his own sex, a model of the other, and a model of the relationship between them. The partnership in procreation continues even after the children are grown because then they (the parents) are needed to help them (the children) establish their own new families.[26]

Fertility - Understanding our fertility helps us make healthier choices in relationships and reproductive choices. If you do not have a committed relationship with someone and you are having sex with them, you risk choosing them as the mother or father for your child. Take a moment and think about the power of your fertility.

Becoming A Whole Person
In Lesson One, we looked at a human being as a whole person. In our human nature, our body, mind, will, emotions, as well as social and moral senses, function together to make us who we are.

Popular media often judges the whole person by appearance or by what one can or cannot do. **This view of the whole person reduces us as objects to be used.**[27]

Some believe a person isn't a person until after birth. Even Dr. Seuss has an opinion!

Do you know what he says about a person? *In Horton Hears a Who*, Horton says, *"And, even though you can't hear them or see them at all, a person's a person, no matter how small."* [28]

Ppt. 2, Lesson 2

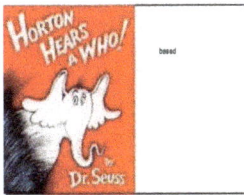

Now we will watch a video to see a powerful visualization through ultrasound, showing human development from conception to birth.

Ppt. 3, Lesson 2
(**Play Video** – *9 Months in the Womb*, PregnancyChat.com[29] (4:36 min.)
https://youtu.be/WH9ZJu4wRUE

(Display Touch of Life Fetal Models 30 12-30 weeks gestation)

Let's take a few minutes to pass around these fetal models so you can get a more real tangible picture of prenatal development.

Question:
How and when do you define a whole person? Do you define a person before or after birth; by what one can do or what one looks like?

Ppt. 4, Lesson 2

Question?

How and when do you define a whole person?

Do you define them before or after birth; by what one can do or what one looks like?

Giving Value & Dignity To A Human Person

Activity $20 Bill:
I have a $20 bill in my hand, how many of you would like to have it? (Now wad the bill up, throw it on the floor and stomp on it pressing it hard and saying it isn't any good.) Now how many of you still want it? You probably still want it because you know, regardless of what it looks like, it still has value.

According to the Universal Declaration of Human Rights[31] *set forth in 1948 by the General Assembly of the United Nations*, human beings (men & women) are so valuable that we come with "human rights" already attached.

"Human rights" mean that we owe each other respect and protection for our unique intrinsic dignity (who we are on the inside). These rights are called "inalienable rights" because they cannot be removed from us by anyone.

Article No. 3 of the declaration says everyone has the right to life, liberty, and security of persons.[32]
Or . . . life, liberty, (our freedom), and the pursuit of happiness as we say it in the United States. The pursuit of happiness or the security of persons would be like owning cars or homes.

Question: Which of these are most important to you? Why would life be most important? If we don't have life, we cannot have other rights.

The Dignity of the Human Person
Ppt. 5, Lesson 2

The Dignity of a Human Person

- inner core of intrinsic value, the natural innate (inborn) desire to give and receive unconditional love

- made to fulfill his/her highest potential by developing a sense of identity, self-worth, personal insight, meaning and purpose

- self-control or the power over our mind, heart and will that is kept by the ability to tell right from wrong

- ability to surrender to self and become emotionally committed to a loving and lasting monogamous relationship in marriage and express the same unconditional love we have received

(Read slide) A human person has at their inner core an intrinsic value, the natural innate (inborn) **desire to give and receive unconditional love**.

Made to fulfill his/her highest potential by developing a sense of identity, self-worth, personal insight, meaning and purpose

As humans, we have self-control or the power over our mind, heart, and will that is kept by the ability to tell right from wrong.

(Slide con't) As the human person develops into a mature adult, this gives us the ability to surrender to self and become emotionally committed to a loving and lasting monogamous relationship in marriage and express the same unconditional love we have received.

Outward or Inward Dignity

Regardless of a human being's looks, conditions, what they can or cannot do, humans have value based on their inalienable rights and this value or right to life, liberty and security of the person cannot be removed

Giving someone value or dignity (respect) based on what they look like, what they own, or what they can do is valuing a person **outwardly or extrinsically**. The popular media equates the outward conditions, physical looks, performances, success, power, and even sex with value. None of these will ever really satisfy the deep longing, innate desire for peace, happiness, and unconditional love.

Also, if we only look at a person's behavior and success in what they can do, we miss seeing who they really are on the inside.
As human beings, we are whole persons. Our sexuality is an integral part of our unique and personal inner core. In this inner core is the natural human desire to give and receive unconditional love.

When we look at the whole person and one's ability to give and receive love in the world, we see the person's **inward or intrinsic dignity.**

Ppt. 6, Lesson 2

Outward & Inward Dignity

Outward or Extrinsic Dignity:
Looks, material things, behaviors & successes in what one can do

Inward or Intrinsic Dignity
Desire & capacity to give and receive unconditional love

We are all here to help you know that you have intrinsic value as a human person. Looks, material things or success, does not matter and doesn't define you as a whole person. **You are a unique, valuable, whole human person! There is not another one in the world just like you!**

Remember we all long and desire good things. How we value ourselves will determine how we value others.

No one wants to be used and abused. Even a young toddler knows the difference when someone intentionally trips them or when it is an accident. If you have been hurt by someone, tell a trusted adult, your parents, a teacher, or a counselor. We all long to be respected and valued as a whole person. We long for peace and happiness. We have the capability as whole persons to give and receive unconditional love.

The following definition of unconditional love can be useful in treating others with respect, dignity as humans, and applying intrinsic value to our relationships.

Ppt. 7, Lesson 2

UNCONDITIONAL LOVE
is patient and kind.
It does not envy.
It does not boast.
It is not proud.
It is not rude.
It is not self-seeking.
It is not easily angered.
It keeps no record of wrongs.
Unconditional love does not delight in evil.
And it rejoices in the truth.
It always protects, always hopes, and always perseveres.
Unconditional love never fails.

Remember the definition of Sexual Health is. . . *living out one's sexuality in true freedom with integrity, purity, and unconditional love. It is making healthy choices in relationships, love, and responsibility that will affect the whole person present and future and the heritage passed on to future generations.*
Take a moment and think about any areas you would like to change in your life.

Let's review the five aspects of the whole person on five fingers. Show your neighbor your respect for their dignity as a whole person by giving them a high five!

Ppt. 8, Lesson 2

Person
As a
WHOLE

Y♀U

Are Uniquely Different!

Are You Dying to Have Sex or Saving Sex for One Monogamous Bond?

A human person has the ability to reason and make choices. Our bodies can be trained to live with sexual health. In this lesson, the difference between love and lust is explored along with the medical facts of sexually transmitted diseases/STDs/STIs. The majority of this lesson is based on students exploring the consequences of STDs and the results of having sex outside of marriage or one monogamous bond. Since contraceptives do not always provide protection against STDs, have negative physical consequences of artificial hormones on the body, and the negative emotional consequences on relationships, this curriculum explains but does not promote their use. However, it will be explained that there are other medical reasons for their use. Students will understand the current epidemic of STDs/STIs in the United States today. Students are encouraged to examine the data and to consider the long-term outcomes of the potential risks of multiple sexual partners as compared to a monogamous relationship as the safest way to avoid STDs/STIs.

Future Plans:
When students begin to grasp the value of sexual health, their lives can change positively, impacting their future choices, relationships, and academic goals. Also, allowing students to evaluate this program is important. The evaluation gives them a voice to agree or disagree to remain pure until marriage or sex with one monogamous partner. And, it gives them the opportunity to recommend the program to others. Students receive a business size commitment card that can serve them in the future as a contract to live out the values and concepts of sexual health. Teachers have the opportunity to reward those who learn the definition of sexual health and also remind the students about the services of the local counseling center.

Objectives:
1) Students will be able to identify the increase in sexual activity and the negative habits in relationships through the use of contraceptives reducing love to lust with at least 75% participation from the class.

2) With at least 90% participation and thorough discussion of the STD activity, students will be able to compare and contrast the risks of contracting multiple STDs and the negative consequences of multiple sexual partners to the benefits of one monogamous relationship with unconditional love.

3) At the end of the lesson, students will be able to describe the difference between bacterial, viral, and parasitic STDs and the negative physical and emotional consequences to their sexual health.

4) Students will apply the knowledge they have learned by writing the definition of sexual health with 100% accuracy, with at least 90% of students participating. Those who are successful will be rewarded by their name going into a drawing for a free prize.

5. Students will give feedback through evaluating the program concerning their plans to apply the benefits of living out the definition of sexual health and recommending the program to others with 90% student participation.

Lesson Plan

Permission is given by the authors to insert appropriate substitutes for video clip(s), handouts, or slides based on current stats/population/culture, maintaining the meaning and values of the original content.

Activity: **STDs are a Sticky Business** (need a piece of chewing gum for each student)
(For Question Pg. 33 need multi-colored Gummy Worm)
Video/slides on STDs (Allow approx. 25-30 min.)

Lecture & Discussion

Handouts: Student's Evaluation of the Program (Appendix, Lesson 3), SH Commitment Cards (may be ordered from Concepts of Truth, Inc. 1.870.238.4329)

PowerPoints (Ppts): (1) Wedding Picture (2) Concepts of Truth's Definition of Sexual Health (3) Question (4) Choices, Choices (5) Planning My Future (6) The Life Wheel: 7 Aspects (7) The Life Wheel: 7 Aspects continued (8) Life Wheel (9) Question (10) Actions Do Speak Louder Than Words (11) Awareness Video about the Dangers of Sexting (12) The Dangers of Sexting (13) Love vs. Lust (14) Question (15) Sexual Exposure Chart (16) Sexual Health Is . . .

Lesson 3 Classroom
Are You Dying to Have Sex or Saving Sex for One Monogamous Bond? Future Plans?

Introduction:

How many of you want to get married or save sex for one monogamous bond? We can find our prince or princess. (Show first slide of wedding pic.) This is Maggie and Justin. They saved sex until they got married. Maggie says Justin was a "real Prince" in that he respected and valued her enough that he wanted to save their sexual relationship for marriage. People have the ability to reason and make choices. Choices produce healthy or unhealthy consequences. Our bodies can be trained to live with sexual health.

Ppt. 1, Lesson 3

We know sexual health is...*living out one's sexuality in true freedom with integrity, purity, and unconditional love. It is making healthy choices in relationships, love, and responsibility that will affect the whole person present and future, and the heritage passed on to future generations.* So, let's review and say it together:

Ppt 2, Lesson 3

(Concepts of Truth's Definition)
Sexual Health is.........
living out one's sexuality in true freedom with integrity, purity, and unconditional love.
It is making healthy choices in relationships, love, and responsibility that will affect the whole person present and future, and the heritage passed on to future generations.

Question:

Answer this question to yourself. Do you feel like you are living out your sexuality in true freedom? Y/ N

If you are not and you are in a relationship where you are being used or abused, please reach out to an adult who you can trust. Talk with your parents, school counselor, teacher, or us. We want to help you value yourself.

We sometimes make decisions that are unhealthy because we don't have direction. In some cases, bad things happened when we were younger and were unable to control the situation. If this has happened to you, it is not your fault. For those things over which we had no control, there is help to heal the pain. If you need help, please call our confidential helpline at **1.866.482.LIFE.**

Ppt. 3, Lesson 3

Question ?

Answer this question to yourself, Do you feel like you are living out your Sexuality in true freedom? Y/N

1.866.482.LIFE

For those things we can control, it is good to recognize how important our decisions are today. Remember, you can choose relationships that reflect unconditional love and pass on a healthy heritage for your children.

Question:

Ask yourself this question, What choices am I making?

Ppt 4, Lesson 3

Remember, our choices affect our whole person. They affect us intellectually, morally, emotionally, physically, and socially. Look at the next PowerPoint revealing some good consequences based on whether one chooses to save sex for one monogamous partner. Our choices concerning our attitudes and behaviors about sex can be healthy, and the consequences can be good!

Ppt. 5, Lesson 3

Our Choices Affect The Whole Person

Good consequences can be yours if you save sex for one monogamous bond.

Intellectual
Graduate
High Self-Esteem

Emotional
No Regret
Happy
Have a Life
Confident
Stable
Without Hurt

Moral
Right Thinking
Peace
Connected

Physical
Future Children
Healthy Relationships
Sexual Health
Marriage

Social
Financially Secure
Good Reputation
Respect Boyfriend/girlfriend
Married with Unitive Intimacy

Our choices of our behaviors say a lot about who we are, how we view ourselves, what we think of others, and what we believe. According to psychologist, Revel Miller, Ph.D, in the article, *The Life Wheel: 7 Aspects of Who You Are:* [33]

Take time to read PowerPoints 6 and 7 for *The Life Wheel: 7 Aspects of Who You Are* described.

Ppt. 6, Lesson 3

- Behavior encompasses how you act or perform from second to second throughout your lifetime.
- Your behavior is judged by others and may present itself within normal limits or it may be quite unusual.
- You are responsive to your environment and you influence things and others around you through your behavior.
- You may also refer to your behavior as how you handle or manage yourself, your comportment, mannerisms, style, routine and speech.
- You usually behave in accordance with your belief systems, your cultural norms and your ethical standards.
- **As a human, you express and regulate your emotions and ideas, your values and morals, your beliefs and ideals through your bodily action. (Bold emphasis added by Concepts of Truth, Inc.)**

Ppt. 7, Lesson 3
The Life Wheel: 7 Aspects of Who You Are:

- You can be highly controlled or very spontaneous, flexible or inflexible, and open or closed.
- At times, you are aware of how your behaviors affect others. At other times you may be oblivious of your impact on other people.
- Sometimes you care how you impress others. While at other times, you may show no concern about how others regard your behaviors.
- You are probably comfortable expressing some behaviors that seem to be automatic and habitual.
- Other behaviors may be new. So, you experiment and try new behaviors to test how they work and feel.
- Some behaviors take a lot of practice and talent to acquire. Some are easier to learn. With others, you may find yourself to be quite clumsy and awkward.

So, think back to our 5 aspects of the human person and look at this diagram of **The Life Wheel** [34] to see how all of the aspects of who we are work together. Our sexuality is an integral part of our entire being and our unique and personal inner core. In this inner core is the natural human desire to give and receive unconditional love. So, in the middle of this wheel, we see our **SELF**, our inner core, which is integrated (hold up hand and wiggle fingers) into our entire being. We express ourselves, our sexuality through our behaviors.

Ppt. 8, Lesson 3

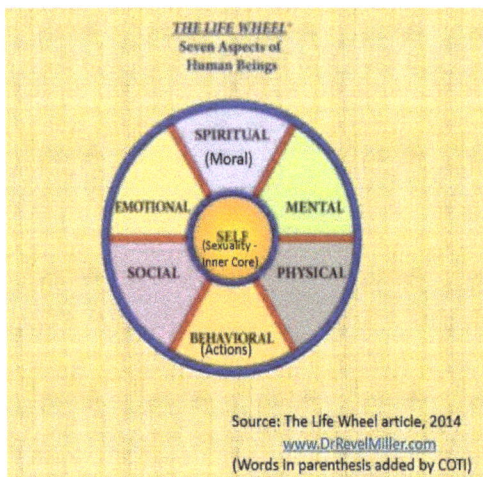

THE LIFE WHEEL
Seven Aspects of Human Beings

SPIRITUAL (Moral)
EMOTIONAL
MENTAL
SELF (Sexuality - Inner Core)
SOCIAL
PHYSICAL
BEHAVIORAL (Actions)

Source: The Life Wheel article, 2014
www.DrRevelMiller.com
(Words in parenthesis added by COTI)

Ppt. 9, Lesson 3

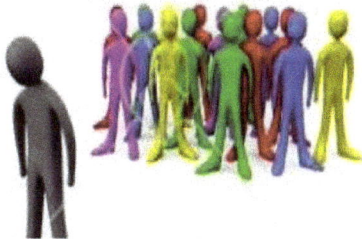

Sometimes we say one thing, but our body language and behaviors are saying something else and we send the wrong message.

Question:

How would you change your behavior to portray a better picture of who you really are? Remember the gummy worm?

Think about this: (**Show another multi-color Gummy Worm**) Worms are attractive to fish. What can you do to make your behaviors more attractive to others so they will see your intrinsic value? **Take a moment and think about this important question.**

Ppt. 10, Lesson 3

Question:

How would you change your behavior to portray a better picture of who you really are?

Remember the

gummy worm?

This Photo by Unknown Author is licensed under CC BY-SA

What can you do to make your behaviors more attractive to others so they will see your intrinsic value?

Lust or Love?

If sexual health includes true freedom, integrity, purity, and unconditional love, then sex is good and pure and right. If sex is good, why do people accept being used as a tool, hooking up or being a friend with benefits?

People who do not know how to live with sexual health experience second rate sex in relationships outside of marriage. This causes pain and brokenness and people get hurt. Today, many adolescents get involved in the dangerous activity of sexting. Let's watch this video.

Ppt. 11, Lesson 3

(Play Video: Awareness Video About the Dangers of Sexting)[35]
https://youtu.be/MoRtLk1xihY

And, please pay attention to this news clip.

Ppt. 12, Lesson 3

(Play Video: The Dangers of Sexting)[36]
https://youtu.be/w13y5KojR3k

Now is a good time to start making some decisions about how you will let yourself be treated.

Lesson 1 explained that real love can only come through making a total gift of self in the commitment to one monogamous partner and hopefully for a lifetime. Love involves a sacrifice of self, respecting the other person, not using them and surrendering yourself to an unconditional love that can be modeled in your relationship as an example to others.

Hookups or friends with benefits are based on lust, extrinsic value, and is temporary. This reduces the person to an object to be used. Compare the characteristics of lust and love in the following Powerpoint list as we read them together:

Ppt. 13, Lust vs. Love

Lust
• Temporary
• Based on fantasy
• Shallow
• Selfish
• Sudden
• Can't wait to get
• Focuses on external looks
• Full of emotion

Love
• Enduring
• Based on reality
• Deep
• Unselfish
• Gradual
• Can't wait to give
• Focuses on internal character
• Full of devotion

Question:
Ppt. 14, Lesson

Ask yourself this question:
Am I being used and abused in my relationship/s?

Contraceptives (Birth Control)

Most people will agree, since contraceptives (birth control pills or condoms) are more available, sexual activity has increased among teens. Most adolescents are trying to prevent pregnancy by using birth control. They are not thinking about preventing sexually transmitted disease. Research tells us that birth control does not always protect against STDs.

Activity: STDs are Sticky Business - (You will need a piece of gum for each student who participates.)

Please take a piece of gum and begin to chew it, but make sure you save your wrapper. (Allow a short time for students to enjoy the gum.) Now, please put your gum back in the wrapper and place it in the collection basket. (Teacher mixes up the pieces of gum and offers them back to the students.) Now, you may find your piece of gum and chew it again. (The responses will probably be "Yuck", "no way", etc.)

Most of you do not want to attempt to find your gum because you are afraid that you will get someone else's, correct? I do not think anyone wants to put gum in his or her mouth knowing it has been in someone else's mouth, correct? Then why would a person have sex with more than one partner if he or she were not absolutely sure that person had never had sex with anyone else? Research tells us that STDs (sexually transmitted diseases) are passed from person to person in bodily fluids or by body contact. If this activity is symbolic of having sex with more than one partner and you choose to take the gum back again, would anyone be disease free?

Just like refusing to take the gum back, everyone has a choice to save sex for one monogamous partner. Let's look at this sexual exposure chart and see how a person's risk of contracting a sexually transmitted disease is exponentially multiplied when having multiple sex partners[37]

Ppt. 15, Lesson 3   **Ppt. 16, Lesson 3**

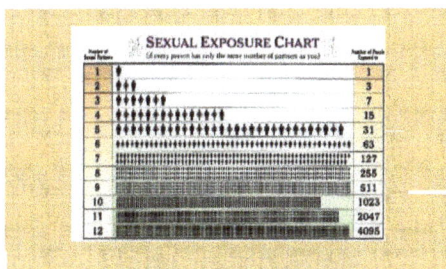

(Concepts of Truth's Definition)
Sexual Health is.........
living out one's sexuality in true freedom with integrity, purity and unconditional love.

It is making healthy choices in relationships, love and responsibility that will affect the whole person present and future and the heritage passed on to future generations.

Good to review Concepts of Truth's Sexual Health definition here
(Note: 45-minute class time ends here.)

Lesson 3 Classroom – Second Half
(Note: 45-minute class time begins here.)

Page | 36

Today is our last meeting. We have thoroughly enjoyed getting to know you. We have one more important set of slides and videos to show you that will educate you on STDs.

What does STD stand for? (Sexually Transmitted Disease) or sometimes you may hear them called STIs. The following slides/video will cover some of the most common bacterial, viral, and parasitic sexually transmitted diseases.

Show the STD Slides/Video[38] https://www.youtube.com/watch?v=bhC5hRm673M (Allow 20 min.)
Take time to discuss briefly and ask Review Questions with Key. (Appendix, Lesson 3)

Questions:
What new information did you learn about STDs today? How will this new information affect your future sexual decisions and/or those of your current and future relationships? Do you think a person might lie to a potential sexual partner about their past? Would this affect your decision to engage in sex with someone if you could not be sure of their sexual history?

Having sex outside of marriage or one monogamous bond increases your risk of not only pregnancy and STDs, but also all kinds of other relationship problems. In the last few lessons, hopefully you have learned the definition of sexual health and that living out that definition will help you in all aspects of your whole person - intellectual, moral, emotional, physical, and social.

You have been taught the development, intrinsic dignity, and value of a human person; and you have had the opportunity to explore how your choices affect your future and your children's future. You have the right to life, liberty, and the security of person. Sex was created for procreation, bonding, completeness, unitive intimacy, and yes, it is pleasurable! It is healthier to save sex for marriage.

Handout: Evaluation Handout & Definition of Sexual Health (Appendix Lesson 3)

Please use the evaluation form to write the definition of Sexual Health to the best of your memory. Even if you think you only know part of it, you may know more than others in the class.

Your input on the evaluation form will be helpful in many ways. Rating the program allows us to have feedback on the program's effectiveness.

You do not have to put your entire name on the evaluation, just your first name and last initial so we can sort it for the drawing. Also, please indicate if you are male or female in the top right corner. It will be exciting to see how many will stay committed to living out your sexuality with integrity, purity, and unconditional love.

Let's see how many got all of the following definition correct.

Ppt 16, Lesson 3

(Concepts of Truth's Definition)
Sexual Health is.........

living out one's sexuality in true freedom with integrity, purity and unconditional love.

It is making healthy choices in relationships, love and responsibility that will affect the whole person present and future and the heritage passed on to future generations.

Y U

Handout: SH Commitment Cards

The SH commitment card has a place for your signature on the back if you will commit to living out the definition of sexual health. This will be a keepsake for you and also the card can be a daily reminder of your promise.

A contact phone number and web information are listed on the card. We would love for you to keep in touch with us or come by our office anytime.

In closing, we thank you for participating and being attentive. We want to encourage you to never forget what you have learned and to always strive to live out the definition of sexual health.

Also, please tell someone else about what you have learned in these lessons. **Remember, there is no one else in the world just like you. You are a whole person and you are valuable!**

Let's see who won the gift!

APPENDIX

STD Slides/Video Review
 Questions

1. Do people infected with Chlamydia always know it?
   Yes   No

2. All STDs can be cured.

   True   False

3. A person ALWAYS has symptoms with Gonorrhea.

   True   False

4. Gonorrhea can be spread to the throat by oral sexual contact.

   True   False

5. Once a person is infected with Herpes, he/she will be infected for life.

   True   False

6. People who have NO signs or symptoms may transmit herpes.

   True   False

7. Three kinds of STDS are

   a. Viral, Penile, Anal

   b. Bacterial, Viral, Parasitic

   c. Chronic, Lifelong, Incurable

   d. None of the above

8. HIV can be passed on through touching, hugging, kissing, or other
   casual contact.

   True   False

9. There is no cure for HIV/AIDS.

   True   False

10. Which of the following STDS/STIS are caused by a parasite?

   a. Scabies

   b. HPV

   c. Pubic Lice

   d. both a & c

STD Slides/Video Review Questions Answer Key

1. No
2. False
3. False
4. True
5. True
6. True
7. B
8. False
9. True
10. D

Date_____   Please circle M for male or F for female   M   F

Name of School_____   Class Period_____

Concepts of Sexual Health Sex & You! Jr High Evaluation Sheet

(Please write the definition of Sexual Health as presented in the program)

Sexual Health is ....

Rate the *Concepts of Sexual Health Sex & You!* Program - Please rate how the program was or was not beneficial to you. Please take the time to give us your rating on the following questions. Put an X in the box under your choices. Thanks and have a healthy life.

Place an X under 1 item only for each statement.	Agree	Disagree	Undecided
1. I plan to stay pure, save sex until marriage or change my sexual behaviors based on what I learned from *Concepts of Sexual Health Sex & You!*			
2. After completing *Concepts of Sexual Health Sex & You!*, I would recommend the program to others.			
3. I understand more about my sexual health as a whole person because of the *Concepts of Sexual Health Sex & You!* program.			

Please tell us something that you learned from *Concepts of Sexual Health Sex & You!*

Rate the Presenters

Rate the Presenters - Put an X under your choice for each comment.	Fair	Good	Excellent
1. Presenters were knowledgeable about the Sexual Health information shared.			
2. Presenters shared the information in an understandable manner.			
3. Presenters had a caring attitude toward the students.			

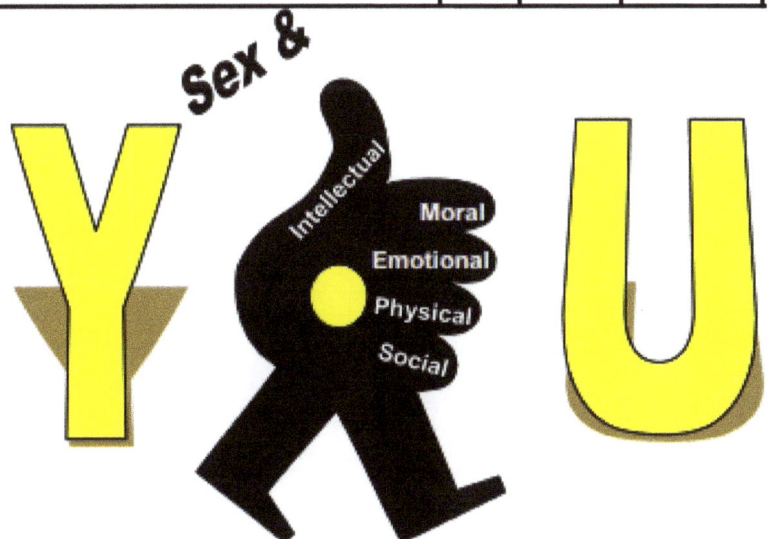

Concepts of Sexual Health

SH Evaluation Totals Page

Name of Presenter (s) _____ Date_____

Phone _____   Email _____

Name of School _____   Class Period _____

Teachers   _____

<u>Total Students in Class</u>   Class Start Date _____

End Date _____

Boys   _____

Girls   _____

Total Students   _____

% of Students Committed To Purity (No. 1 under Rate SH Program)

% Agree		Agree	Disagree	Undecided
_____	Boys			
_____	Girls			
	Totals			
Total % of students Committed to Purity		%		

% of Students that would Recommend SH (No. 2 under Rate SH Program)

% Agree		Agree	Disagree	Undecided
_____	Boys			
_____	Girls			
	Totals			
Total % of students Recommended SH		%		

_____
Signature of Presenter

Available for future presentations?   Y N   If yes, days available M T W T F

(Circle)

Notes

[1] Arkansas 2013 State Health Profile http://www.cdc.gov/nchhstp/stateprofiles/pdf/arkansas_profile.pdf

[2] Onalee McGraw, Teaching the Whole Person about Love, Sex and Marriage, Educating for Character in the Common World of our Homes, Schools and Communities (Educational Guidance Institute, Front Royal, Virginia, 2004), 43.

[3] Concepts of Sexual Health, Sex & You Curriculum, (Concepts of Truth, Inc., 2010, 2015, 2018 Revised)

Arkansas Grade 6 Student Expectations	Arkansas Grade 7 Student Expectations	Arkansas Grade 8 Student Expectations
H.W. 6.6.1 — Les 2	H.W. 6.7.2 — Les 2	H.W. 6.8.2 — Les 2
H.W. 6.6.2 — Les 1	H.W. 7.7.4 — Les 3	H.W. 7.8.4 — Les 3
H.W. 7.6.1 — Les 3	H.W. 7.7.5 — Les 3	H.W. 7.85 — Les 3
H.W. 7.6.4 — Les 3	H.W. 7.7.6 — Les 3	H.W. 8.8.1 — Les 1
H.W. 7.6.5 — Les 3	HW. 8.7.1 — Les 1	H.W. 9.8.1 — Les 1
H.W. 9.6.1 — Les 1	H.W. 9.7.1 — Les 1	H.W. 9.8.4 — Les 1
H.W. 9.6.3 — Les 1	H.W. 9.7.4 — Les 1	H.W. 11.8.3 — Les 1
H.W. 9.6.4 — Les 1	H.W. 11.7.3 — Les 1	H.W. 11.8.12 —Les 3
H.W. 11.6.3 — Les 1	H.W. 11.7.12 — Les 3	
H.W. 11.6.11 — Les 3		
H.W. 11.6.12 — Les 3		

[4] McDowell, Josh, "Rules Without Relationships=Rebellion", Frequently Asked Questions, http://www.josh.org

[5] Onalee McGraw, Teaching the Whole Person about Love, Sex and Marriage, Educating for Character in the Common World of our Homes, Schools and Communities (Educational Guidance Institute, Front Royal, Virginia, 2004), 52.

[6] Ibid, 51.

[7] Ibid.

[8] WHO World Health Organization definition of Sexual Health, 2002

[9] Onalee McGraw, Teaching the Whole Person about Love, Sex and Marriage, Educating for Character in the Common World of our Homes, Schools and Communities (Educational Guidance Institute, Front Royal, Virginia, 2004),12

[10] Ibid.

[11] J. Budziszewski, "What We Lose When We Forget What Sex Is For", (Fellowship of St. James, 2005), http://www.Touchstonemag.com/archives/print.php

[12] "Hardwired to Connect: The New Scientific Case for Authoritative Communities. (New York: Institute for American Values, 2003), 31-32

[13] Bill & Pam Farrel, Men Are Like Waffles, Women Are Like Spaghetti (Harvest House Publishers, 2007)

[14] Budziszewski, J., "What We Lose When We Forget What Sex Is For", (Fellowship of St. James, 2005), http://www.Touchstonemag.com/archives/print.php

[15] Universal Declaration of Human Rights",1948, http://www.un.org/en/documents/udhr/

16 Budziszewski, J., "What We Lose When We Forget What Sex Is For", (Fellowship of St. James, 2005), http://www.Touchstonemag.com/archives/print.php

[17] Ibid.

18 Ibid.

[19] Developing the Whole Person Video, 2018, https://www.youtube.com/watch?v=VDFonhPChmY

[20] Barbara Wilson, The Invisible Bond, (Colorado Springs, Co: Multnomah Books, 2006), 30-33

[21] "Science of Sex", http://www.humanlife.org/onlinecitations.php

[22] Ibid.

[23]Joe S. McIlhaney, Jr., M.D., Freda McKissic Bush, M.D., Hooked, New Science On How Casual Sex Is Affecting Our Children, (Northfield Publishing, Chicago, Illinois, 2008)

R. E. Rector, K. A. Johnson, and L. R. Noyes, "Sexually Active Teenagers Are More Likely to Be Depressed and to Attempt Suicide," Washington D.C.: A report from the Heritage Center for Data Analysis, "The Heritage Foundation. Publication CDA03-04, June 2 (2005)

Barnet et al., 2004; Breheny & Stephens, 2007; Forum on Child & Family Statistics, 2007; Hofferth et al., 2001; Hoffman, 2006, Pregnancy and Childbearing Among U.S. Teens http://www.planned-parenthoodnj.org/library/topic/family_planning/teen_pregnancy

 L. Coley and P. L. Chase-Lansdale, "Adolescent Pregnancy and Parenthood: Recent Evidence and Future Directions," American Psychologist 53, no. 2 (1998): 152-166.

Nicole M. Else-Quest, Janet Shibley Hyde, and John D. DeLamater, "Context Counts: Long-Term Sequelae of Pre-marital Intercourse or Abstinence", The Journal of Sex Research, Vol. 42, 2005

[24]Lee A. Lillard, Michael J. Brien, and Linda J. Eaite, "Pre-Marital Cohabitation and Subsequent Marital Dissolution: Is It Self-Selection?" Demography 32 (1995): 437-458.

Elizabeth Thomsom and Ugo Collela, "Cohabitation and Marital Stability: Quality or Commitment?" Journal of Marriage and the Family 54 (1992): 259-268.

G. K. Rhoades, S. M. Stanley, H. J. Markman, "Pre-engagement cohabitation and gender asymmetry in marital commitment," Journal of Family Psychology (Dec. 2006): 20 (4):553-60.

C. T. Kenney, S. S. McLanahan, "Why are cohabitating relationships more violent than marriages?" Demography (Feb. 2006): 43 (1): 127-40

[24a] Ibid.

[25] 9 Months in the Womb, PregnancyChat.com, 2014, https://www.youtube.com/watch?v=WH9ZJu4wRUE

[26] Harvard University Study "Dad is Destiny", U.S. News and World Report, (Feb. 27, 1995)

[27]J. Budziszewski, "What We Lose When We Forget What Sex Is For", (Fellowship of St. James, 2005), http:// www.Touchstonemag.com/archives/print.php

[28] Horton Hears a Who!, (2008), http://en.wikiquote.org/wiki/Horton_Hears_a_Who!_(film)

[29] 9 Months in the Womb, PregnancyChat.com, 2014, https://www.youtube.com/watch?v=WH9ZJu4wRUE

[30] Touch of Life Fetal Models, (Heritage House'76, Inc., Snowflake, Arizona)

[31]"Universal Declaration of Human Rights", 1948, http://www.un.org/en/documents/udhr/

[32] Ibid.

[33] Revel Miller, Ph.D, in the article, The Life Wheel: 7 Aspects of Who You Are (©2014)

[34] Ibid.

[35] Video, Awareness About the Dangers of Sexting, Warwickshire Police, 2018, https://www.youtube.com/watch?v=MoRtLk1xihY

[36] Video, The Dangers of Sexting, WXYZ-TV Channel 7 Detroit, 2014, https://www.youtube.com/watch?v=w13y5KojR3k

[37] Sexual Exposure Chart, wvdhhr.org/.../edresources/sexual_exposure_chart.pdf

[38] STD Slides/Video, (Concepts of Truth International, 2015, 2018 Revised, 2019 Modified for Jr. High)

www.ingramcontent.com/pod-product-compliance
Lightning Source LLC
Chambersburg PA
CBHW051557030426
42334CB00034B/3471